ICME-13 Topical Surveys

Series editor

Gabriele Kaiser, Faculty of Education, University of Hamburg, Hamburg, Germany

More information about this series at http://www.springer.com/series/14352

Christine Suurtamm · Denisse R. Thompson
Rae Young Kim · Leonora Diaz Moreno
Nathalie Sayac · Stanislaw Schukajlow
Edward Silver · Stefan Ufer
Pauline Vos

Assessment in Mathematics Education

Large-Scale Assessment and Classroom Assessment

Christine Suurtamm
Faculty of Education
University of Ottawa
Ottawa, ON
Canada

Denisse R. Thompson
Department of Teaching and Learning
University of South Florida
Tampa, FL
USA

Rae Young Kim
Mathematics Education
Ewha Womans University
Seoul
Korea (Republic of)

Leonora Diaz Moreno
Departamento de Mathématicas
Universidad de Valparaiso
Valparaíso
Chile

Nathalie Sayac
Val de Marne
University Paris Est Créteil
Créteil Cedex
France

Stanislaw Schukajlow
Instituts für Didaktik der Mathematik und der
  Informatik
Westfälische Wilhelms-Universität Münster
Münster, Nordrhein-Westfalen
Germany

Edward Silver
University of Michigan
Ann Arbor, MI
USA

Stefan Ufer
Lehrstuhl für Didaktik der Mathematik
Ludwig Maximilians-Universität München
München
Germany

Pauline Vos
University of Agder
Kristiansand
Norway

ISSN 2366-5947
ICME-13 Topical Surveys
ISBN 978-3-319-32393-0
DOI 10.1007/978-3-319-32394-7

ISSN 2366-5955   (electronic)

ISBN 978-3-319-32394-7   (eBook)

Library of Congress Control Number: 2016936284

Printed on acid-free paper

This Springer imprint is published by Springer Nature
The registered company is Springer International Publishing AG Switzerland

Main Topics Found in This ICME-13 Topical Survey

- Purposes, Traditions, and Principles of Mathematics Assessment
- Design of Assessment Tasks in Mathematics Education
- Mathematics Classroom Assessment in Action
- Interactions of Large-scale and Classroom Assessment in Mathematics Education
- Enhancing Sound Mathematics Assessment Knowledge and Practices.

Acknowledgments

We acknowledge the valuable input of Karin Brodie, University of Witwatersrand, South Africa and Co-chair of TSG 40, in the development of this topical survey.

Contents

Assessment in Mathematics Education

1 Introduction

Although classroom teachers have long used various forms of assessment to monitor their students' mathematical learning and inform their future instruction, increasingly external assessments are being used by policy makers throughout the world to gauge the mathematical knowledge of a country's students and sometimes to compare that knowledge to the knowledge of students in other countries. As a result, external assessments often influence the instructional practices of classroom teachers. The importance given to assessment by many stakeholders makes assessment a topic of interest to educators at many levels.

Because we believe those interested in large-scale assessment as well as classroom assessment have much to offer each other, the Co-chairs and committee members of the two Topic Study Groups (TSG) at ICME-13 focused on assessment in mathematics education, TSG 39: *Large-Scale Assessment and Testing in Mathematics Education*, and TSG 40: *Classroom Assessment for Mathematics Learning*, chose to work together to develop this volume, to consider research and discussions that might overlap as well as those that are specific to either classroom or large-scale assessment in mathematics education. By developing this survey of the field and considering the work being done in assessment throughout the world, we hope to provide a common foundation on which discussions can build and we offer potential questions for further research and discussion.

This volume draws on research to discuss these topics and highlights some of the differences in terms of challenges, issues, constraints, and affordances that accompany large-scale and classroom assessment in mathematics education as well as some of the commonalities. We recognize there are some strong differences between the traditions, practices, purposes, and issues involved in these two forms of assessment. But we also propose that there are overlapping areas that warrant discussion, such as assessment item and task design, and connections and implications for professional knowledge and practice. This volume provides the

© The Author(s) 2016
C. Suurtamm et al., *Assessment in Mathematics Education*,
ICME-13 Topical Surveys, DOI 10.1007/978-3-319-32394-7_1

opportunity to discuss both large-scale and classroom assessment in mathematics education as well as their interactions through the following main themes:

- Purposes, Traditions, and Principles of Mathematics Assessment
- Design of Assessment Tasks in Mathematics Education
- Mathematics Classroom Assessment in Action
- Interactions of Large-scale and Classroom Assessment in Mathematics Education
- Enhancing Sound Mathematics Assessment Knowledge and Practices.

2 State of the Art

This main body of the survey provides an overview of current research on a variety of topics related to both large-scale and classroom assessment in mathematics education. First, the purposes, traditions, and principles of assessment are considered with particular attention to those common to all assessment and those more connected with either classroom or large-scale assessment. Assessment design in mathematics education based on sound assessment principles is discussed with large-scale and classroom assessment being differentiated, but also with a discussion of how the design principles overlap. We then discuss mathematics classroom assessment and provide some specific examples of assessment strategies. The impact of large-scale assessment in mathematics education on curriculum, policy, instruction, and classroom assessment follows. We conclude by discussing the challenges that teachers face, as well as ways to support them. In each section, we pose problems worthy of continued or future research. We hope this discussion will provide common language for the TSGs in assessment as well as lay out some issues that we hope to discuss.

2.1 Purposes, Traditions, and Principles of Assessment

This section sets the stage for the rest of the volume as it discusses purposes, principles, goals and traditions of assessment in mathematics education, with a view to discussing commonalities as well as differences between large-scale and classroom assessment. We recognize and discuss some strong differences between the traditions, practices, purposes, and issues involved in these two forms of assessment.

2.1.1 Purposes of Assessment

Assessment has been used for multiple purposes, such as providing student grades, national accountability, system monitoring, resource allocation within a district, student placement or monitoring, determining interventions, improving teaching and learning, or providing individual feedback to students and their parents/guardians (Newton 2007). Purpose is important. Claims that are made should be different, depending on the goals and design of the assessment activity.

Large-scale and classroom assessment serve different purposes and have different goals. Large-scale assessment informs systems. It is often used for system monitoring, to evaluate programs, or to make student placements. In many jurisdictions around the world, students are assessed in mathematics using some form of large-scale assessment that may take the form of national, state, or provincial assessments but could also take the form of international assessments. For accountability purposes, such large-scale assessments of mathematics are used to monitor educational systems and increasingly play a prominent role in the lives of students and teachers as graduation or grade promotion often depend on students' test results. Teachers are sometimes evaluated based in part on how well their students perform on such assessments (Wilson and Kenney 2003).

Classroom assessment gathers information and provides feedback to support individual student learning (De Lange 2007; National Research Council [NRC] 2001b) and improvements in teaching practice. Classroom assessment usually uses a range of teacher-selected or teacher-made assessments that are most effective when closely aligned with what and how the students have been learning (Baird et al. 2014). Current perspectives in classroom assessment encourage the use of a range of assessment strategies, tools, and formats, providing multiple opportunities for students to demonstrate their learning, making strong use of formative feedback on a timely and regular basis, and including students in the assessment process (e.g., Brookhart 2003; Klenowski 2009; National Council of Teachers of Mathematics [NCTM] 2014).

Whether talking about large-scale or classroom assessment, attention must be paid to the purpose of the assessment so the results of the assessment are interpreted and used appropriately for that purpose. The results of the assessment should be used for the purposes for which the assessment was designed and the inferences need to be appropriate to the assessment design (Koch 2013; Messick 1989; Rankin 2015). However, we recognize that purposes of assessment are sometimes blurred. For instance, teachers often use summative classroom assessments for formative purposes when they use the summative results to understand students' misconceptions and design future instruction accordingly, or when they use tasks from large-scale assessments for instructional purposes (e.g., Kenney and Silver 1993; Parke et al. 2003). Additionally, in some parts of the world data from large-scale assessments are provided to teachers to better understand teaching and learning in their classrooms (see for example, Boudett et al. 2008; Boudett and Steele 2007).

2.1.2 Assessment Traditions

Large-scale assessment and classroom assessment have different traditions, having been influenced in different ways by learning theories and perspectives (Glaser and

Silver 1994). Large-scale assessment traditionally comes from a psychometric/ measurement perspective, and is primarily concerned with scores of groups or individuals, rather than examining students' thinking and communication processes. A psychometric perspective is concerned with reliably measuring the outcome of learning, rather than the learning itself (Baird et al. 2014). The types of formats traditionally used in large-scale assessment are mathematics problems that quite often lead to a single, correct answer (Van den Heuvel-Panhuizen and Becker 2003). Some might see these types of questions as more aligned with a behaviourist or cognitivist perspective as they typically focus on independent components of knowledge (Scherrer 2015). A focus on problems graded for the one right answer is sometimes in conflict with classroom assessments that encourage a range of responses and provide opportunities for students to demonstrate their reasoning and creativity, and work is being done to examine large-scale assessment items that encourage a range of responses (see for example Schukajlow et al. 2015a, b).

Current approaches to classroom assessment have shifted from a view of assessment as a series of events that objectively measure the acquisition of knowledge toward a view of assessment as a social practice that provides continual insights and information to support student learning and influence teacher practice. These views draw on cognitive, constructivist, and sociocultural views of learning (Gipps 1994; Lund 2008; Shepard 2000, 2001). Gipps (1994) suggested that the dominant forms of large-scale assessment did not seem to have a good fit with constructivist theories, yet classroom assessment, particularly formative assessment, did. Further work has moved towards socio-cultural theories as a way of theorizing work in classroom assessment (e.g., Black and Wiliam 2006; Pryor and Crossouard 2008) as well as understanding the role context plays in international assessment results (e.g., Vos 2005).

2.1.3 Common Principles

There are certain principles that apply to both large-scale and classroom assessment. Although assessment may be conducted for many reasons, such as reporting on students' achievement or monitoring the effectiveness of an instructional program, several suggest that the central purpose of assessment, classroom or large-scale, should be to support and enhance student learning (Joint Committee on Standards for Educational Evaluation 2003; Wiliam 2007). Even though the *Assessment Standards for School Mathematics* from the National Council of Teachers of Mathematics in the USA (NCTM 1995) are now more than 20 years old, the principles they articulate of ensuring that assessments contain high quality *mathematics*, that they enhance student *learning*, that they reflect and support *equitable* practices, that they are *open* and transparent, that *inferences* made from assessments are appropriate to the assessment purpose, and that the assessment, along with the curriculum and instruction, form a *coherent* whole are all still valid standards or principles for sound large-scale and classroom assessments in mathematics education.

Assessment should reflect the mathematics that is important to learn and the mathematics that is valued. This means that both large-scale and classroom assessment should take into account not only content but also mathematical practices, processes, proficiencies, or competencies (NCTM 1995, 2014; Pellegrino et al. 2001; Swan and Burkhardt 2012). Consideration should be given as to whether and how tasks assess the complex nature of mathematics and the curriculum or standards that are being assessed. In both large-scale and classroom assessment, assessment design should value problem solving, modeling, and reasoning. The types of activities that occur in instruction should be reflected in assessment. As noted by Baird et al., "assessments define what counts as valuable learning and assign credit accordingly" (2014, p. 21), and thus, assessments play a major role in proscribing what occurs in the classroom in countries with accountability policies connected to high-stakes assessments. That is, in many countries the *assessed curriculum* typically has a major influence on the *enacted curriculum* in the classroom. Furthermore, assessments should provide opportunities for all students to demonstrate their mathematical learning and be responsive to the diversity of learners (Joint Committee on Standards for Educational Evaluation 2003; Klieme et al. 2004; Klinger et al. 2015; NCTM 1995).

In considering whether and how the complex nature of mathematics is represented in assessments, it is necessary to consider the formats of classroom and large-scale assessment and how well they achieve the fundamental principles of assessment. For instance, the 1995 *Assessment Standards for School Mathematics* suggested that assessments should provide evidence to enable educators "(1) to examine the effects of the tasks, discourse, and learning environment on students' mathematical knowledge, skills, and dispositions; (2) to make instruction more responsive to students' needs; and (3) to ensure that every student is gaining mathematical power" (NCTM, p. 45). These three goals relate to the purposes and uses for both large-scale and classroom assessment in mathematics education. The first is often a goal of large-scale assessment as well as the assessments used by mathematics teachers. The latter two goals should likely underlie the design of classroom assessments so that teachers are better able to prepare students for national examinations or to use the evidence from monitoring assessments to enhance and extend their students' mathematical understanding by informing their instructional moves. Classroom teachers would often like to use the evidence from large-scale assessments to inform goals two and three and there are successful examples of such practices (Boudett et al. 2008; Boudett and Steele 2007; Brodie 2013) where learning about the complexity of different tasks or acquiring knowledge about the strengths and weaknesses of one's own instruction can inform practice. However, this is still a challenge in many parts of the world given the nature of the tasks on large-scale assessments, the type of feedback provided from those assessments, and the timeliness of that feedback.

In our examination of the principles and goals of assessment in mathematics education from across the globe, we looked at the goals from both large-scale and

Table 1.1 Samples of assessment goals from different perspectives

Designers' goals for high-stakes assessments (Swan and Burkhardt 2012)	Goals for effective formative assessment in Norway (Baird et al. 2014)	Goals for competency models (Klieme et al. 2004)	Goals for assessment based on purpose (Wiliam 2007)
• Measure performance over a variety of types of mathematical tasks • Operationalize objectives for performance in ways that both teachers and students can understand • Identify patterns of classroom instruction (teaching and activities) that would be representative of the majority of classrooms in the system in which students are assessed (adapted from p. 4)	• Students should know what they need to learn • Feedback provided to students should provide information about the quality of their work • Feedback should give insight on how to improve performance • Students should be involved in their own learning through activities such as self-assessment (adapted from pp. 37–38)	• Students use their abilities within the domain being assessed • Students access or acquire knowledge in that domain • Students understand important relationships • Students choose relevant action • Students apply acquired skills to perform the relevant actions • Students gather experience through the assessment opportunities • Cognitions that accompany actions motivate students to act appropriately (adapted from p. 67)	• To evaluate mathematical programs to assess their quality • To determine a student's mathematical achievement • To assess learning to support and inform future instruction (adapted from p. 1056)

formative assessments articulated by educators, assessment designers, or national curriculum documents. Table 1.1 provides a sample of assessment goals that are taken from different perspectives and assessment purposes.

Looking across the columns of the table, there are similarities in the goals for assessment that should inform the design of tasks, whether the tasks are for formative assessment, for competency models, or for high-stakes assessment; in fact, they typically interact with each other. Specifically, teachers and students need to know what is expected which implies that tasks need to align with patterns of instruction, tasks need to provide opportunities for students to engage in performance that will activate their knowledge and elicit appropriate evidence of learning, the assessment should represent what is important to know and to learn, and when feedback is provided it needs to contain enough information so that students can improve their knowledge and make forward progress. These common goals bear many similarities to the 1995 NCTM principles previously discussed.

2.1.4 Suggestions for Future Work

As we move forward in our discussions about the purposes, principles, and traditions of assessment, we might want to consider the following issues:

- How do we ensure that the primary purpose of assessment, to improve student learning of mathematics, is at the forefront of our assessment practices?
- How do we ensure that important mathematical practices/processes are evident in assessments?
- How do we ensure equitable assessment practices? What do these look like?
- How do teachers negotiate the different purposes of classroom and large-scale assessment?
- In what ways are classroom practices influenced by the demands of large-scale assessment?

2.2 Design of Assessment Tasks

This section explores assessment design as it relates to two broad issues:

- The development of assessment tasks that reflect the complexity of mathematical thinking, problem solving and other important competencies.
- The design of alternative modes of assessment in mathematics (e.g., online, investigations, various forms of formative assessment, etc.) (Suurtamm and Neubrand 2015, p. 562).

In so doing, we consider different examples of what we know about the design of assessments and assessment tasks. An extensive discussion of design issues of mathematics tasks can be found in the study volume for ICMI 22 on Mathematics Education and Task Design (Watson and Ohtani 2015).

2.2.1 Design Principles in Large-Scale Assessments

This section presents several models of large-scale assessment design. The models are not meant to be exhaustive but to be representative of some of the approaches that might be used in designing large-scale assessments.

Designing from a psychometric tradition. In comparing and contrasting assessment design from didactic and psychometric models, Van den Heuvel-Panhuizen and Becker (2003) note that large-scale assessments typically require tasks that have a single, correct answer. Because of the closed nature of the task, the answer is all that is assessed, and different methods or strategies to obtain a solution are not important. They summarize design assumptions from a psychometric tradition outlined by Osterlind (1998):

- *Unidimensionality*: Each test item should focus on assessing a single objective or ability.
- *Local independence*: Each test item is independent of every other item, with no hints or expectations that solutions to one item will affect performance on another item.
- *Item characteristic curve*: If items are valid for a particular objective, students of low ability should have a low probability of success on the item.
- *Non-ambiguity*: Items should be stated in such a manner that the beginning portion of the item leads the student to the single correct response (adapted from pp. 700–701).

As Van den Heuvel-Panhuizen and Becker (2003) note, these assumptions are based on beliefs about mathematics problems that most mathematics educators would dispute:

1. "Mathematics problems always have only one correct answer.
2. The correct answer can always be determined.
3. All the needed data should be provided to students.
4. Good mathematics problems should be locally independent.
5. Knowledge not yet taught cannot be assessed.
6. Mathematics problems should be solved in exactly one way.
7. The answer to a problem is the only indicator of a student's achievement level" (p. 705).

From the above, it is clear that designing assessments and tasks according to the assumptions outlined by Osterlind presents a challenge in reflecting current goals in mathematics education that encourage students to demonstrate their thinking, work with real-world messy or ill-structured problems, or solve problems from more than one perspective or that have more than one answer. Psychometric models are often criticized as presenting a narrow view of mathematics.

In the last decade, large-scale international assessment design has begun to consider assessment items that challenge this traditional approach. For instance, some of the recent PISA assessments have attempted to support multiple solutions (Schukajlow et al. 2015a, b). Development continues to address mathematical processes as well as mathematical content.

Designing to integrate content knowledge and mathematical processes. Large-scale assessments are not likely to disappear from the assessment scene in many countries because policy makers view them as a means to monitor the educational system or compare students' performance within or across countries. In other cases, they might function as ways to determine access to higher levels of education. Given the constraint of timed environments in which such assessments occur, Swan and Burkhardt (2012) highlight design principles to ensure high-quality assessments that reinforce the goals for mathematics outlined in the curriculum documents of many countries. They present the following principles for assessment task design:

- Tasks should present a balanced view of the curriculum in terms of all aspects of performance that the curriculum wants to encourage.
- Tasks should have "face validity" and should be worthwhile and of interest to students.
- Tasks should be appropriate for the purpose and should integrate processes and practices rather than attempt to assess those separately from content.
- Tasks should be accessible (have multiple entry points) to students with a range of ability and performance levels, while still providing opportunities for challenge.
- Tasks should provide opportunities to demonstrate chains of reasoning and receive credit for such reasoning, even if the final result contains errors.
- Tasks should use authentic rather than contrived contexts, at times with incomplete or extraneous data.
- Tasks should provide opportunities for students to determine what strategy they want to use in order to pursue a solution.
- Tasks should be transparent enough that students know what types of responses will be acceptable (adapted from p. 7).

Swan and Burkhardt remark about challenges in designing tasks according to these design principles: such tasks are longer than typical tasks in their statement on the page, take more time to complete, and have a higher cognitive load because they are more complex, unfamiliar, and have greater technical demand. When collaborative work is assessed, one needs to consider procedures for assessing the individual students' contribution to the group's workflow and/or products (presentations, posters, etc.) (Webb 1993). Others have raised language issues with these more challenging types of tasks, particularly when the language of assessment may be different from students' mother tongue, thus raising issues of equity within assessment (e.g., Dyrvold et al. 2015; Levin and Shohamy 2008; Ufer et al. 2013). Issues about the layout of a task and the amount of scaffolding that a task might require have encouraged or required task designers to use an engineering design approach, in which they try tasks with students from the target populations, revise and try again, in a continuous cycle of design and refinement.

Designing from a competency model. "Competency models serve … to describe the learning outcomes expected of students of given ages in specific subjects … [and] map out possible 'routes of knowledge and skills' … [so that they] provide a framework for operationalisations of educational goals" (Klieme et al. 2004, p. 64). As Klieme et al. note, before competency models can be used to build assessments, there is a need for empirical study to determine various levels of competence, which are "the cognitive abilities and skills possessed by or able to be learned by individuals that enable them to solve particular problems, as well as the motivational, volitional and social readiness and capacity to use the solutions successfully and responsibly in variable situations" (p. 65).

Within this perspective, competence links knowledge with skills along a spectrum of performance through a variety of tasks that get at more than just factual knowledge. "The basic idea of this model is that a person's mathematical competence can be described with reference to tasks that can be assigned a specific level of difficulty. Individuals at the lowest competency level are able to retrieve and apply their arithmetical knowledge directly. Those at the highest competency level, in contrast, are capable of complex modelling and mathematical argumentation" (Klieme et al. p. 68). These issues suggest the need for measures that are multidimensional rather than unidimensional to reflect competence at meeting various targets and to identify educational standards that students have met and areas where more work is needed (Klieme et al. 2004). Defining competencies in stages, such as learning trajectories, are often hard to operationalize into test design. To be useful, one needs to make inferences from performance to competency, which can be difficult and is a typical critique of this model.

In some European countries, including Switzerland, Belgium and France, competence is connected to unfamiliarity of tasks (Beckers 2002; Perrenoud 1997; Rey et al. 2003). Concerning mathematical competencies, the following is considered as a competency: "a capacity to act in an operational way faced with a mathematical task, which may be unfamiliar, based on knowledge autonomously mobilised by the student" (Sayac and Grapin 2015). Thus in some cases, unfamiliarity is considered a component of a task's difficulty.

In developing an actual assessment instrument to be used across schools, Klieme et al. mention the need to utilize several technical aspects of test development:

- Determining whether the performance will be judged in terms of norm referencing or criterion referencing;
- Determining whether competence will be reported in terms of a single level or across different components of overall competence;
- Determining whether test takers all complete the same items or complete different samples of items;
- Determining whether one test is used for all competence levels or whether test items might be individualized in some way (Klieme et al. 2004, adapted from p. 76).

Others have raised additional technical issues in the design of assessments. For instance, Vos and Kuiper (2003) observed that the order of test items influences results, showing that more difficult items negatively affect achievement on subsequent items. Thus, studies on the design of a large-scale assessment are often conducted to ensure that the effects of the order of test items are minimized. For example, researchers develop different test booklets with varied difficulties within test booklets and the same test item occupies different positions in different versions of the test (OECD 2009).

2.2.2 Design Principles for Classroom Assessment Tasks

Several of the design principles outlined in Sect. 2.2.1 for large-scale assessments of mathematics could be useful for the design of many instructional and assessment tasks used by classroom teachers. However, the more intimate nature of the classroom and the close relationship of teachers to their students provide opportunities for classroom tasks to be designed using some principles that may not be practical in large-scale assessment environments. We now consider some models within this context, again using the principles within these models to highlight issues and possibilities rather than in any attempt to be exhaustive.

A didactic model of assessment. A didactic model of assessment design is based on work from the Freudenthal Institute and the notion that mathematics is a human activity in which students need to make meaning for themselves. Designing assessment tasks according to these principles can potentially be quite beneficial for classroom teachers by providing opportunities for students to demonstrate and share their thinking, which can then serve as the basis for rich classroom discussion. Through such activities, teachers are able to determine appropriate and useful feedback that can move students forward in their learning and that can inform the path toward future instruction, both essential elements of formative assessment.

Van den Heuvel-Panhuizen and Becker (2003) suggest that assessment tasks might include problems designed with the following principles in mind:

- Tasks have multiple solutions so that students can make choices and use their natural thinking and reasoning abilities. Multiple solutions could involve both tasks with multiple pathways to a single solution as well as multiple solutions.
- Tasks might be dependent, that is, tasks might be paired or contain multiple parts where a solution on an earlier part is used to make progress on a subsequent part or subsequent problem. One power of including dependent tasks "is that the assessment reveals whether the students have insight into the relationship between the two problems, and whether they can make use of it" (p. 709).
- Tasks where what is of interest is the solution strategy rather than the actual answer. Teachers are able to consider strategies used to identify situations in which students recognize and use relationships to find solutions in an efficient manner and those in which students may obtain a solution but from a more convoluted approach (Van den Heuvel-Panhuizen and Becker 2003, adapted from pp. 706–711).

Van den Heuvel-Panhuizen and Becker note that tasks designed with these principles in mind tend to be rich tasks that provide opportunities for students to engage with problems of interest to them, take ownership of their learning, and provide insight into thinking that can further classroom instruction and discourse to move students forward mathematically.

Using disciplinary tasks. Cheang et al. (2012) describe a project in Singapore to use disciplinary (or contextual) tasks as a means of making assessment practices an integral part of mathematics instruction. These contextual tasks are designed so that they assess a variety of mathematical competencies. Specifically, Cheang and his colleagues suggest that these tasks incorporate the following:

- Demonstration of an understanding of the problem and how to elicit information from the problem;
- Ability to complete appropriate computations;
- Inductive and deductive reasoning as applicable;
- Communication using relevant representations (e.g., tables, graphs, equations);
- Ability to use mathematics to solve real-world problems.

When designing a set of tasks for use as part of classroom assessment, teachers can analyse which of these design principles are evident in each task or sub-task. If certain principles are not adequately evident, then assessment tasks can be modified so that students are exposed to a balanced set of tasks. They can also consider formats other than written tests, such as projects, group work, or oral presentations for classroom assessment.

2.2.3 Suggestions for Future Work

Much work is still needed to ensure that widely used large-scale assessments are designed according to sound principles so that the positive influences of such assessments on classroom instruction and assessment can be fully realized. Some questions for future research and discussion at the conference might be:

- How might we work with policy makers to design large-scale assessments in line with validity, reliability and relevance (De Ketele 1989)?
- How do we help policy makers and the general public understand the value of assessments that ask for more than a single correct answer or that require students to write extended responses, particularly when such assessments may cost more to administer and evaluate?
- What support is needed to help students value assessment tasks with extended responses and use them to gauge their own learning?
- What support is needed to help teachers develop and use assessments as described in Sect. 2.2.2 in their own classrooms on a regular basis?
- How does teachers' content domain knowledge interact with their ability to use data from large-scale assessments or what support might they need to use such data?
- How does the use of technology influence the design of assessment items? What are the affordances of technology? What are the constraints?
- How can students be prepared for unexpected or unfamiliar tasks?
- How can collaborative tasks be incorporated into assessments?
- What training or skills do teachers need to develop or be able to analyse/critique assessments?

- What issues do teachers face and how might they be overcome when large-scale assessments are delivered through interfaces or formats quite different from those used during regular classroom assessment?
- How might research from assessments or assessment design help construct new knowledge in mathematics education as is occurring in France (e.g., Bodin 1997)?

2.3 Classroom Assessment in Action

The preceding sections have outlined a number of principles that should be considered when looking at designing an assessment plan for a classroom. These principles are designed to provide opportunities for students to engage in meaningful mathematics and demonstrate their thinking, both to provide evidence of their learning and to inform teachers about future instruction. Current perspectives in mathematics education encourage teachers to engage students in developing both content and process (i.e., reasoning and proof, representations, connections, communication, problem solving) knowledge and to ensure that students develop robust mathematical proficiency consisting of procedural fluency, conceptual understanding, adaptive reasoning, productive dispositions, and strategic competence (National Research Council 2001a). Teachers have been prompted by educational policy, teacher journals, and professional development initiatives to incorporate new assessment practices in their classrooms in order to develop a better understanding of student thinking and to provide appropriate feedback (Wiliam 2015). Shifts to a broad range of classroom assessment practices are encouraged by both the current classroom assessment literature (e.g., Gardner 2006; Stobart 2008) and by recent thinking and research in mathematics education (e.g., NCTM 1995, 2000; Wiliam 2007).

In this section, we discuss principles and issues that are fundamental to classroom assessment, and then describe some specific examples from around the globe that teachers have used to respond to the principles of classroom assessment.

2.3.1 Assessment as an On-going Process in the Classroom

One test alone cannot adequately assess the complex nature of students' mathematical thinking. Rather, different types of assessment are required to assess complex processes such as problem solving, justifying or proving solutions, or connecting mathematical representations. As a way to listen and respond to student thinking, teachers are encouraged to use a variety of formats for assessment, such as conferencing, observation, or performance tasks [i.e., "any learning activity or assessment that asks students to *perform* to demonstrate their knowledge, understanding and proficiency" as defined by McTighe (2015)]. Many of these practices are part of formative assessment.

In this document, we use the term formative assessment to indicate informal assessments that teachers might do as part of daily instruction as well as more formal classroom assessments used to assess the current state of students' knowledge. Then teachers use that information to provide feedback to students about their own learning and to plan future instruction. This perspective seems to be in tune with perspectives from other educators, as noted in the representative definitions provided here:

- "Assessment for learning [formative assessment] is part of everyday practice by students, teachers and peers that seeks, reflects upon and responds to information from dialogue, demonstration and observation in ways that enhance ongoing learning" (Klenowski 2009, as cited in Baird et al. p. 42).
- "Practice in a classroom is formative to the extent that evidence about student achievement is elicited, interpreted, and used by teachers, learners, or their peers to make decisions about the next steps in instruction that are likely to be better, or better founded, than the decisions they would have taken in the absence of the evidence that was elicited" (Black and Wiliam 2009, as cited in Wiliam 2011b, p. 11).

Both definitions are supported by the following five strategies or principles to guide teachers in formative assessment as outlined by Leahy et al. (2005) as cited in Wiliam (2011a):

- Teachers should clarify learning outcomes and conditions for success and then share them with students.
- Teachers should engage students in classroom activities that provide evidence of learning.
- Teachers should provide feedback to help students make progress.
- Students should be resources for each other.
- Students should own their learning (adapted from p. 46).

In work with groups of teachers throughout England, Black et al. (2004) identified four practices that teachers could use to engage in continuous and formative assessment as part of classroom instruction:

- questioning with appropriate wait time so that students have an opportunity to think about an acceptable response and so that worthwhile, rather than superficial, questions are asked;
- providing feedback without necessarily attaching a grade because attaching a grade can have a negative impact on students' perceptions of their work and cause them to ignore helpful feedback that informs students about what aspects of their response are strong and what might need improvement;
- helping students learn to peer and self assess, both of which are critical components in assisting students to take ownership of their own learning; and
- using summative assessments in formative ways by helping students learn to develop potential questions for a summative assessment and determine what acceptable responses might entail.

The practices identified by Black et al. resonate with the work of Suurtamm, Koch, and Arden who note that reform of assessment practices has meant a changing view of assessment from "a view of assessment as a series of events that objectively measure the acquisition of knowledge toward a view of assessment as a social practice that provides continual insights and information to support student learning and influence teacher practice" (2010, p. 400). This suggests that assessment is an integral part of instruction and that students have multiple and on-going opportunities to demonstrate their thinking to the teacher and their peers, so that this evidence of knowledge (or the lack thereof) can be used to move learning forward. Through study of the assessment practices of a group of Canadian teachers, Suurtamm et al. articulated several practices reflective of this broad perspective in relation to assessment:

- teachers used various forms of assessment;
- instruction and assessment were seamlessly integrated as teachers built assessment into instructional tasks so they were constantly assessing students' thinking;
- teachers valued and assessed complex thinking through problem solving; and
- teachers used assessment to support their students' learning by providing students with appropriate feedback about their thinking, by helping them learn to self-assess, and by using students' thinking to guide their classroom instruction (adapted from Suurtamm et al. 2010, pp. 412–413).

Viewing assessment as an ongoing and natural part of an instructional plan has also been evident in research exploring assessment practices of Finnish teachers. Teachers viewed assessment as "continuous and automatic" so that, as they planned instruction, they considered "what kind of information is needed, how and when it is collected and how it is used" (Krzywacki et al. 2012, p. 6666).

2.3.2   A Sampling of Classroom Assessment Strategies and Tasks

As noted in the previous section, assessment should be an on-going endeavour that is an integral part of classroom instruction. In Sects. 2.1 and 2.2 we outlined some general principles to guide sound assessment design. In this section, we report on a sampling of classroom assessment strategies that have been reported to help teachers develop assessments that make student understanding explicit and provide evidence of robust understanding.

Kim and Lehrer (2015) use a learning progression oriented assessment system to aid teachers in developing tasks that help students make conceptual progress in a particular content domain. Their work entails developing construct maps that are the outcomes of a learning progression, assessment items intended to generate the types of reasoning identified in the construct maps, scoring exemplars, and then lesson plans with contexts that enable students to engage with representations of the mathematics.

Thompson and Kaur (2011) as well as Bleiler and Thompson (2012/2013) advocate for a multi-dimensional approach to assessing students' understanding, building from curriculum work originating with the University of Chicago School Mathematics Project in the USA. They suggest that, for any content topic, teachers might consider tasks that assess understanding of that content from four dimensions: Skills (S), which deal with algorithms and procedures, Properties (P) which deal with underlying principles, Uses (U) which focus on applications, and Representations (R) which deal with diagrams, pictures, or other visual representations of the concepts. This SPUR approach to understanding and assessing helps ensure that teachers not only teach from a balanced perspective but assess from that balanced perspective as well.

For example, consider the content domain of decimal multiplication. Figure 1.1 contains a sample problem for this concept within each dimension. If students can solve a skill problem but not represent the concept visually, what does that say about the robustness of students' understanding and what are the implications for teachers' instructional decisions? Thompson and Kaur share results from an international study with grade 5 students in the USA and Singapore that suggest students' proficiency is often different across these four dimensions, perhaps because of differential emphasis in the curriculum as well as in instruction.

Toh et al. (2011) discuss the use of a practical worksheet to assess mathematical problem-solving within Singapore classrooms, with a particular focus on the processes used when solving problems rather than solely on the final solution. Based on the problem solving work of Pólya and Schoenfeld, the practical worksheet has students make explicit statements that indicate how they understand the problem, what plans they develop and implement in an attempt to solve the problem, what key points and detailed steps were taken at various decision points along the plan, and how they checked their solution and expanded the problem. Thus, the extensive work students complete to make their thinking visible to the teacher, to their peers, and to themselves provides critical information that can be used to identify misconceptions in thinking so that appropriate actions can be implemented that will

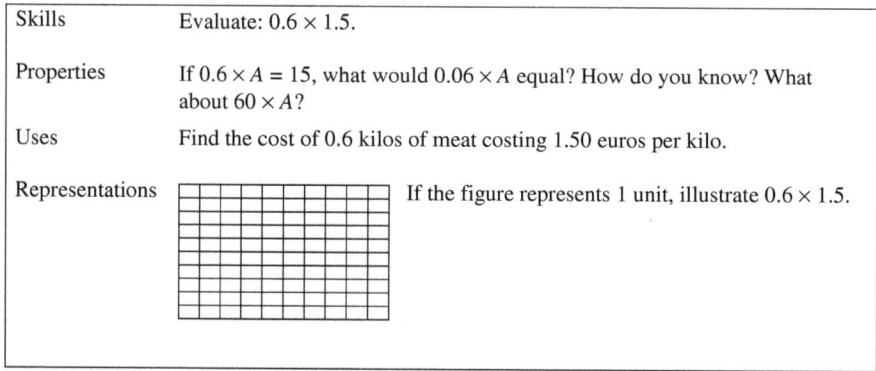

Fig. 1.1 Decimal multiplication assessed using the SPUR approach

help students move forward. As students become familiar with the scoring rubric associated with the practical worksheet, they are able to use it to monitor and assess their own understanding and problem-solving endeavours.

Young-Loveridge and Bicknell (2015) report the use of culturally-sensitive task-based interviews to understand the mathematical proficiency of a diverse group of young children in New Zealand. The researchers used contexts that would be familiar to the children being interviewed because of their cultural significance, such as monkeys with bunches of bananas. By attending to contexts that were less likely to disadvantage children, the researchers were able to explore children's understanding of concepts, such as multiplication and division, which the children could reason about through the contextual cues although they had not formally studied the concepts.

Elrod and Strayer (2015) describe the use of a rubric to assess university mathematics students' ability to engage in whole-class and small group discourse while also communicating about their problem solving in written form. Their rubric assesses discourse and written work for problem solving, reasoning and proof, representation, communication, connections, and productive disposition with ratings as *above the standard, at the standard, below the standard*, or *unacceptable*. The rubric became a tool for teachers to use as they monitored students working in groups but was also a tool for students to use in assessing the work of their peers. The researchers discuss how the rubric became a means to reference sociocultural norms expected within the classroom, and thus, became an integral part of teachers' assessment practices of students. Smit and Birri (2014) describe somewhat similar work with rubrics in primary classrooms in Switzerland and found that the work with rubrics helped both students and teachers understand competencies required on national standards and was a means to provide beneficial feedback.

The strategies described briefly in the previous paragraphs are just a sample of alternative assessment strategies that classroom teachers can use on a regular basis to engage students in the types of mathematical problems and discourse that have the potential to elicit student thinking and inform instruction. Numerous other strategies have been identified in the literature, such as concept mapping (Jin and Wong 2011), journal writing (Kaur and Chan 2011), or exit slips, learning logs, and "find the errors and fix them" (Wiliam 2011a).

2.3.3   Self Assessment

As indicated in Sect. 2.3.1, one of the foundational principles of formative assessment is to help students become actively engaged in assessing their own learning and taking action on that assessment. As cited by Wiliam (2011a), the Assessment Reform Group in the United Kingdom indicated that, if assessment were to improve learning, then students needed to learn to self-assess in order to make improvements, and that teachers needed to recognize that assessments can influence students' motivation and self-esteem. So, what are some approaches that teachers might use to support students as they learn to self-assess their work as well as assess the work of peers?

Mok (2011) describes the Self-directed Learning Oriented Assessment (SLOA) framework developed and implemented with students in China, Hong Kong, and Macau. The framework consists of three components: Assessment *of* Learning gives the student insight into what has already been learned and what gaps exist between current understandings and intended learning; Assessment *for* Learning provides feedback to help the student move forward in his/her mathematical understanding; and Assessment *as* Learning is designed to help the student learn to self-assess and direct his/her own learning. Mok describes the pedagogical approaches teachers might use within each component, some tools to facilitate those approaches, and some pitfalls to avoid. For instance, in assessment as learning, teachers help students learn to set goals, monitor their program, and to reflect on their learning. Students can complete learning goal surveys with questions such as "Can I reach this goal?" and then with choices such as "absolutely confident, strong confident, rather confident, absolutely not confident" (p. 209). Rather than assume students will develop the ability to assess on their own, students need role modelling and explicit instruction to develop these essential skills.

Fan (2011) suggests three different approaches that teachers can use to implement self-assessment into their instruction. In *structured self-assessment*, teachers have students complete pre-designed assessment survey forms during instruction to gauge how students perceive their understanding or as a summative survey at the end of a unit. Sample prompts might include "This topic is overall easy" or "I can complete homework for this topic most of the time by myself" (p. 281). In *integrated self-assessment*, students might complete a survey that is simply a part of a larger assessment package, with prompts such as "What were the mathematical ideas involved in this problem?" or "What have you learnt from this presentation [after presentation of an extended project, for example]?" (p. 283). In *instructional self-assessment*, teachers embed self-assessment into the typical classroom activities, perhaps even informally or impromptu. Interviews with students in classes that participated in a research study related to these components of self-assessment found that the activities promoted students' self-awareness and metacognition; they learned to be reflective and think deeply about their own learning.

2.3.4 Suggestions for Future Work

It is evident that the current vision of mathematics instruction requires on-going assessment that provides evidence for teachers of student understanding that can inform instruction. But as Swan and Burkhardt (2012) note in their discussion of assessment design and as noted by other researchers in assessment (e.g., Black et al. 2004), change is not easy or swift. Teachers and students need time to experiment with changes, sometimes being successful and other times failing, but learning from those failures and trying again. That is, teachers and students need space within which to become comfortable with new assessment practices and to reflect on their benefits and challenges.

As the community continues the study of assessment in action, some questions to consider and discuss might be the following:

- What concerns about equity come into play with these suggested changes in assessment? How might teachers support students whose cultural background discourages disagreements with the teacher? How do we support teachers and students when cultural backgrounds might suggest more teacher-centered rather than student-centered classrooms?
- What are some of the additional challenges in assessment when hand-held technologies are available (e.g., graphing calculators) or mobile technologies are easily accessible (e.g., smart phones with internet connections)?
- How does technology access influence equity, both within and across countries?

2.4   Interactions of Large-Scale and Classroom Assessment

Assessment is a part of the procedure of making inferences, some of which are about students, some about curricula, and some about instruction (Wiliam 2015).

> Assessment is always a process of reasoning from evidence. By its very nature, moreover, assessment is imprecise to some degree. Assessment results are only estimates of what a person knows and can do (Pellegrino et al. 2001, p. 2).

Assessment results in mathematics education are and have been used in a variety of ways, particularly when we examine the impact of large-scale assessment on policy, curriculum, classroom practice, and individual student's careers. When large-scale assessments focus on monitoring, they are at the system level and some might suggest that there is no direct impact upon teachers and learners. Such large-scale assessments are perceived as having little to say about individuals, because they do not deliver sufficiently reliable results for individuals, but only on higher levels of aggregation. However, in some countries (France, Belgium, Netherlands, Norway, etc.) large-scale assessments include national examinations that all students must take in order to progress to further studies. Such large-scale assessments are exit assessments, whereby students cannot leave secondary school without passing the national exams. The question arises as to what extent are large-scale assessments for accountability for teachers and students used and how might such use influence the nature of classroom instruction?

2.4.1   Assessments Driving Reform

One issue is whether the results of large-scale assessment should drive curriculum and instructional reform (Barnes et al. 2000). Standings on international assessments of mathematics often drive political and educational agendas and there have been several examples of this in situations in recent years (c.f. Klieme et al. 2003). In particular, if countries have nationally organised exit examinations, these may drive (or hinder) reform. Some suggest that externally imposed assessments have been used in attempts to drive reform in some cases (Earl and Torrance 2000; Mathematical Sciences Education Board [MSEB] and National Research Council [NRC] 1993). For instance, in many countries the OECD PISA results have affected curriculum in such a way to focus it more specifically on particular topics, such as problem solving (De Lange 2007). "All of these efforts were based on the idea that assessment could be used to sharpen the focus of teachers by providing a target for their instruction" (Graue and Smith 1996, p. 114).

2.4.2   Impact of Large-Scale Assessment on Classroom Assessment

The nature and design of assessment tasks and the results of assessments often have an enormous influence on the instruction orchestrated by teachers. Swan and Burkhardt (2012) note that the influence of assessments is not just on the content of instruction but on the types of tasks that students often experience (e.g., multiple-choice versus more open or non-routine problems). If students and teachers are judged based on the results of large scale assessments, in particular in countries with national exit examinations, there is a need for students to experience some classroom assessments that are related to the types of tasks used on such assessments so the enacted curriculum is aligned with the assessed curriculum and so that results of the assessments provide useful and reliable evidence to the educational system. Some have questioned whether large-scale assessments that are used for accountability measures can be used or designed to generate information useful for intervention by teachers (Care et al. 2014). Also, as indicated by Swan and Burkhardt and Klieme et al., there is a need for piloting of assessment tasks or empirical study of competence levels before these are used in or applied to large-scale assessments. Piloting provides opportunities for trying tasks in classroom contexts. As Obersteiner et al. (2015) discuss, when items are developed to assess competence at different levels and then students' solutions are analysed for potential misconceptions from a psychological perspective, teachers are able to use competency models in ways that can guide instruction and help students move forward in their mathematical understanding.

   However, there is some caution about the use of large-scale assessment types in classroom assessment. If teachers teach to tests constructed using a psychometric model, they are likely to teach in a superficial manner that has students learn in small chunks with performance demonstrated in isolated and disconnected segments of knowledge. A study by Walcott and colleagues challenges a practice of preparing students for large-scale multiple-choice assessments by having the majority of classroom assessments take a multiple choice format. The study results show that, in fact, the students in classrooms where the teachers are using a variety of assessments seem to have higher achievement (Walcott et al. 2015). Their results do not necessarily demonstrate a causal relationship but provide evidence that a constant use of multiple-choice tests as classroom assessments may not necessarily lead to high achievement. Brunner and colleagues provide further evidence of the effects of teaching to the test. They report that, when students were taught using PISA materials, their performance on the PISA assessment was similar to the control group in the study (Brunner et al. 2007). Others have found that assessments that accompany published materials teachers use in classrooms often fail to address important process goals for mathematics, such as the ability to engage in reasoning and proof, communicate about mathematics, solve non-routine problems, or represent mathematical concepts in multiple ways (e.g., Hunsader et al. 2014; Sears et al. 2015).

2.4.3 "Teaching to the Test"

Many might argue or complain that assessments force too many teachers to *teach to the test*. Some would argue against large-scale assessments driving classroom instruction, as often the nature of the large-scale assessment might narrow the curriculum. Swan and Burkhardt argue that if the assessment is well designed and aligns with curriculum goals then 'teaching to the test' is a positive practice. They suggest:

> To make progress, it must be recognised that high-stakes assessment plays three roles:
>
> A. Measuring performance across the range of task-types used.
> B. Exemplifying performance objectives in an operational form that teachers and students understand.
> C. Determining the pattern of teaching and learning activities in most classrooms (p. 4).

Swan and Burkhardt note that large-scale assessments provide messages to classroom teachers about what is valued and have an impact on both instructional and assessment activities within a classroom. Their work suggests a need to develop assessments that pay attention to their implications and align not only with mathematics content but also with mathematical processes and actions.

However, Swan and Burkhardt note that if the assessments reflect what is important for students to learn about mathematics, both in terms of content and performance, then "*teachers who teach to the test are led to deliver a rich and balanced curriculum*" (2012, p. 5, italics in original). They argue that if an assessment is perfectly balanced and represents what is important to learn, then perhaps 'teaching to the test' can be a good thing. This scheme of 'teaching to the test' has been put forward as a potential tool for innovation and has also been articulated by other scholars (Van den Heuvel-Panhuizen and Becker 2003). As De Lange noted in (1992):

…[throughout the world] the teacher (or school) is judged by how well the students perform on their final exam. This leads to test-oriented learning and teaching. However, if the test is made according to our principles, this disadvantage (test-oriented teaching) will become an advantage. The problem then has shifted to the producers of tests since it is very difficult and time consuming to produce appropriate ones. (as cited in Van den Heuvel-Panhuizen and Becker, p. 691)

An important question, then, is whether the assessment adheres to design principles that engage students with tasks that provide opportunities to engage with important mathematical processes or practices. However, even the best-designed assessments cannot account for issues such as test anxiety. When assessment activities mirror instructional activities, that anxiety might be reduced and such assessments might be a more accurate indicator of student achievement.

2.4.4 Examples of Positive Interaction Between Large-Scale Assessment and Classrooms

If the enacted curriculum of the classroom and the assessed curriculum are to inform each other and to enhance student learning in positive and productive ways, then large-scale external assessments cannot operate in isolation from the classroom. A number of researchers from different parts of the globe have documented the interplay of these two assessment contexts.

Shimizu (2011) discusses bridges within the Japanese educational system between large-scale external assessments and actual classrooms. As in many countries, he notes tensions between the purposes of external assessments and those of classroom assessments and that teachers are concerned about the influences of external assessments on their classroom instructional practices and assessment. To assist teachers in using external assessments to inform their classroom instruction, sample items are released with documentation about the aims of the item as well as student results on the item, and potential lesson plans that would enable such items to be used in the classroom. For instance, on multiple-choice items, information can be provided that suggests the types of thinking in which students may be engaged based on answer choice, thus assisting teachers to understand the nature of misconceptions in students' thinking based upon the incorrect answer selected. As a result, external assessment tasks provide students with opportunities for learning.

Given the expectation that the assessed curriculum reflects what should be taught within the classroom, Shalem et al. (2012) describe the benefits of having South African teachers engage in curriculum mapping of large-scale assessments. The authors acknowledge that assessment developers often engage in such activities, but that classroom teachers rarely do. For a given test item, teachers identified the concept being assessed, identified and justified the relevant assessment standard, and then determined if the necessary content was taught explicitly or not and at what grade. The results of the curriculum mapping helped to highlight discrepancies between the intended curriculum assumed by assessment designers and the actual enacted curriculum of the classroom. Specifically, there were three important outcomes for teachers: teachers developed an understanding of the content assessed at a particular grade and at what cognitive demand; teachers could reflect on their own practice as they identified whether the content was actually taught in class-rooms; and teachers developed a more robust understanding of the curriculum. All three of these outcomes provided an important link between the external assessment and actual classroom practice that could inform future instruction.

Brodie (2013) discusses an extension of the work of Shalem, Sapire, and Huntley in the South African context. In the Data Informed Practice Improvement Project, teachers work in professional learning communities to analyse test items, interview students, map curriculum to assessments, identify concepts underlying errors, and then read and discuss relevant research. Within school based learning communities, teachers use school level data to design and then reflect on lessons to address issues within the data analysis.

As a final representative example of how external assessments might be used to inform classroom instruction and assessment, we consider the work of Paek (2012). Paek suggests that learning trajectories are one means "to make an explicit and direct connection of high-level content standards, what is measured on high-stakes large-scale assessments, and what happens in classrooms" (p. 6712). She argues that development of such trajectories "requires deep understanding of the content, how students learn, and what to do when students are struggling with different concepts" (p. 6712). Knowledge of such trajectories can be used in the development of assessment items but also can help to inform teachers about how students learn particular concepts. In her project, Paek works with researchers, mathematics educators, and national consultants to develop resources for teachers that connect learning trajectories to big ideas within and across grades. It is hoped that by becoming more familiar with learning trajectories, teachers can consider developing classroom assessment items that focus on different aspects of the learning contin-uum or that provide opportunities for transferring of skills or that enable student work to be collected to monitor progress along the continuum.

2.4.5 Making Use of Assessment Results

Making use of assessment data is strongly connected to assessment literacy. Webb (2002) defines assessment literacy as:

> the knowledge of means for assessing what students know and can do, how to interpret the results from these assessments, and how to apply these results to improve student learning and program effectiveness (p. 4).

This definition of assessment literacy can apply to the interpretation of results from both classroom and large-scale assessments. In fact, the challenges of making use of assessment results span both classroom assessment results and large-scale assess-ment results.

Making sense of classroom assessment results is often seen as teachers' work. Wyatt-Smith et al. (2014) acknowledge that, along with the task of designing assessments, teachers need to know how to use assessment evidence to identify the implications for changing teaching. However, interpreting classroom assessment results is also the work of parents, students, and school administrators. In terms of students, including students in the assessment process has been shown to provide them with a stronger sense of what is being assessed, why it is being assessed, and ways they can improve, thus allowing them to make better use of assessment results (Bleiler et al. 2015; Tillema 2014). Developing students' ability to self-assess is an important aspect of this work (Fan 2011). Once students have a strong assessment literacy, they can assist their parents in interpreting assessment results, but this should be supported with solid and frequent communication between schools, teachers, students, and parents in order to enhance the assessment literacy of all who are involved in understanding assessment results.

Making sense of large-scale assessment results is often seen as the purview of many education stakeholders, including teachers, administrators, school district officials, state or provincial policy makers, national policy makers or administrators, as well as the general public. Each group has its own perspective on the results, sometimes multiple contrasting perspectives, which can complicate interpretation. To the extent that large-scale assessment results are useful for more than conversations across constituent groups, the value will come in the application of suitable interpretations to educational policy and practice. The question arises as to what do educators need to know to make productive use of assessment results? The first priority when examining test results is to consider the purpose of the test and to view the results in light of the assessment's purpose. Those viewing the results should also have some assessment literacy to be able to know what they can and cannot infer from the results of the assessment. Rankin (2015) places some of this responsibility on the assessment developer and suggests that, in order to assist educators in making use of assessment results, those responsible for conveying the results should consider how they are organized and what information is provided to those using the results. She suggests that data analysis problems often occur because the data are not presented in an organized and meaningful manner. Her work highlights several common types of errors that are made in the interpretation of assessment results and she cautions educators to pay close attention to the assessment frameworks and literature published in concert with the assessment's results (often on a website). She presents a series of recommendations for which she feels educators should advocate that include improvements to data systems and tools that will facilitate easier and more accurate use of data. See also Boudett et al. (2008) and Boudett and Steele (2007) for other strategies in helping teachers interpret results.

2.4.6   Suggestions for Future Work

There is no doubt that much can be learned from large-scale assessment. Assessments help to highlight achievement gaps, point to areas where adaptations to instruction might be required, and can lead to reforms in curriculum and teaching. International assessments have helped to globalize mathematics education and have created a forum for mathematics educators to share their knowledge and experiences, and work together to resolve common issues.

As shown, large-scale and classroom assessment interact with one another in different ways. Because they rest on similar principles of sound assessment, ultimately having coherence between the two would help to support all students to be successful. Wyatt-Smith et al. (2014) suggest "better education for young people is achievable when educational policy and practice give priority to learning improvement, thereby making assessment for accountability a related, though secondary, concern" (p. 2).

This section has raised many issues for future research. Some questions to consider include:

- What are the challenges and or issues in attempting to bring all stakeholders for assessment to the table—developers of large-scale national or international assessments, textbook publishers or developers, classroom teachers?
- What are issues in attempting to ensure that important mathematical practices/processes are evident in assessments?
- What training or skills do teachers need to develop to be able to analyse/critique assessments?
- Should policy decisions be made based on large-scale assessments, given evidence that some assessments may not reflect students' achievement when the assessment has little impact on students' educational lives (e.g., some students do not give their best effort on the assessment)?

2.5 Enhancing Sound Assessment Knowledge and Practices

Assessing student learning is a fundamental aspect of the work of teaching. Using evidence of student learning and making inferences from that evidence plays a role in every stage of the phases of instruction. This section discusses the challenges teachers face in engaging in assessment practices that help to provide a comprehensive picture of student thinking and learning. It also presents examples of professional learning opportunities that have helped to support teachers as they enhance their assessment practices.

2.5.1 The Assessment Challenge

Assessing students' learning is multi-faceted. The process of making sense of students' mathematical thinking, through student explanations, strategies, and mathematical behaviors, is much more complex than might be anticipated and can often challenge teachers' ways of thinking about mathematics and mathematics teaching and learning (Even 2005; Watson 2006). In many cases, teachers require new assessment practices along with different ways of thinking about teaching and learning. Several researchers have suggested there is variability in the extent to which teachers have implemented innovative assessment practices, often based on teachers' conceptions of assessment and mathematics teaching (e.g., Brookhart 2003; Duncan and Noonan 2007; Krzywacki et al. 2012). Although there might be some resistance to change, even when teachers want to shift their practice and incorporate more current assessment practices, it is difficult to change without many other factors in place, such as changing students', parents', and administrators' views (Marynowski 2015; Wiliam 2015). For instance, while involving students in the assessment process is seen as important (Tillema 2014), Semena and Santos' work (2012) suggests that incorporating students in the assessment process is a very complex task.

2.5.2 Dilemmas Teachers Face

Several researchers have examined the challenges teachers face in shifting or enhancing their practice (c.f. Adler 1998; Silver et al. 2005; Suurtamm and Koch 2014; Tillema and Kremer-Hayon 2005; Windschitl 2002). Using Windschitl's (2002) framework of four categories of dilemmas—conceptual, pedagogical, political, and cultural—to analyze their transcripts, Suurtamm and Koch describe the types of dilemmas teachers face in changing assessment practices, based on their work with 42 teachers meeting on a regular basis over 2 years. They see *conceptual* dilemmas in assessment occur as teachers seek to understand the conceptual underpinnings of assessment and move their thinking from seeing assessment as an event at the end of a unit to assessment as ongoing and embedded in instruction. Adopting new assessment practices may require a deep conceptual shift that takes time and multiple opportunities to dialogue with colleagues and test out assessment ideas in classrooms (Earl and Timperley 2014; Timperley 2014; Webb 2012).

Teachers face *pedagogical* dilemmas as they wrestle with the "how to" of assessment practices in creating and enacting assessment opportunities. These might occur as teachers work on developing a checklist for recording observations, designing a rubric, or asking others for ways to find time to conference with students. For instance, in research in Korea, Kim and colleagues noted pedagogical dilemmas as teachers struggled to create extended constructed response questions (Kim et al. 2012).

Many teachers face *cultural dilemmas*, and often these are seen as the most difficult to solve. They might arise as new assessment practices challenge the established classroom, school, or general culture. For instance, students might confront the teacher with questions if, rather than receiving a grade on a piece of work, the student receives descriptive feedback. Or, cultural dilemmas might arise if the teacher is adopting new assessment practices but is the only one in his or her department to do so and is met with some resistance from others.

Political dilemmas emerge when teachers wrestle with particular national, state, provincial, district, or school policies with respect to assessment. These might arise due to an overzealous focus on large-scale assessment, particular policies with respect to report card grades and comments, or with particular mandated methods of teaching and assessment that do not represent the teachers' views. Engelsen and Smith (2014) provide an example of a political move that may have created such dilemmas for teachers. They examined the implications of national policies to create an Assessment For Learning culture in all schools within a country, and give the instance of Norway. Their examination suggests that this top-down approach through documents and how-to handbooks does not necessarily take into account the necessity for conceptual understanding of sound assessment practices to take hold. They suggest more work with policy makers and administrators in understanding that developing assessment literacy is more important than mandating reforms.

What appears helpful about this parsing out of dilemmas into categories is that different types of dilemmas might require different types of supports. For instance, in many cases, pedagogical dilemmas, the 'how to' of assessment, might be resolved through a workshop or through teachers sharing resources. However, cultural dilemmas are not as easily solved and might require time and focused communication to resolve. Although this parsing out of dilemmas is helpful when looking at supports, it should also be recognized that the dilemmas types interact. In other words, a policy cannot be invoked without understanding that it will have pedagogical, cultural, and conceptual implications and pedagogical resources won't be employed if there is not enough conceptual understanding or cultural shifts.

2.5.3 Supporting Teachers

Some would suggest that adopting and using new assessment practices is a complex process that needs to be well supported, particularly by other colleagues (e.g., Crespo and Rigelman 2015). As Black et al. (2004) note, as part of embedding assessments into instructional practice, teachers first need to reflect on their current practice and then try changes in small steps, much like the "engineering design" approach to assessment development advocated by Swan and Burkhardt (2012). The following presents a sample of models and activities that have been implemented to enhance teachers' understanding of assessment and their assessment practices, and to support them as they negotiate the dilemmas they face in shifting assessment practices.

Hunsader and colleagues have been working with both pre-service and in-service teachers to assist them in becoming critical consumers and developers of classroom assessment tasks (Hunsader et al. 2015a, b). This work has taken place over time, working with teachers in critiquing, adapting, or transforming published assessment tasks so they better address the mathematical actions that students should be taking (e.g., mathematical processes, mathematical practices). This work has reassured teachers that they can adapt and build strong assessment tasks by modifying and adapting readily available resources.

Webb's study (2012) focused on middle school teachers working over a four-year period in a professional learning situation to improve their classroom assessment practice. Teachers engaged in a variety of collaborative activities, such as assessment planning, and analysis of classroom tasks and student work. Teachers tried out the assessment ideas in their classes and brought back those experiences to the group, thus providing authentic classroom connections. The project demonstrated an increase in teachers' use of higher cognitive demand tasks involving problem contexts. Webb suggested that professional learning needed to challenge teachers' prior conceptions of assessment as well as the way they saw mathematics.

Marynowski's study (2015) documented the work of secondary mathematics teachers who were supported by a coach over one year as they worked on co-constructing formative assessment opportunities and practices. One of the positive results was that teachers noticed that students were held more accountable for their learning and were more engaged in the learning process. Teachers appreciated the immediate feedback about student learning but also some teachers noted resistance on the part of students (Marynowski 2015).

Lee et al. (2015) were involved with 6 colleagues in a self-study of a Professional Learning Community (PLC) set within a university course. Within this PLC they sought to refine the assessment practices they were using in their own classrooms. They focused on developing questioning techniques by role playing questioning techniques, refining those techniques, using the techniques with their own students in interviews, and bringing back the results of the conversations to the PLC. The group reported an increased level of awareness of teacher-student dialogue.

Other professional learning activities that had positive impacts on teachers' assessment expertise and practice include lesson study with a focus on assessment (Intanate 2012), working with student work samples from formative assessment opportunities in teachers' classrooms (Dempsey et al. 2015), and shifting from merely feedback to formative feedback for teachers in a Grade 1 classroom (Yamamoto 2012).

There are many common features in the professional learning models above as well as in other studies that point to success. In most cases, the professional learning occurred over an extended period of time from a half year to several years. Many of these models involved components of teachers moving back and forth from the professional learning situation to their own classrooms, thus connecting their learning with authentic classroom situations and situating knowledge in day-to-day practice (Vescio et al. 2008). In some cases, the professional learning was occurring in a pre-service or in-service classroom setting which further provides an opportunity for the mathematics teacher educator to model sound assessment practices. Many models also involved working with a group of teachers, thus creating a community that provided support as teachers tried out new ideas. The Professional Learning Community model provides an environment where the teacher is not working in isolation but is able to share ideas and lean on colleagues to help solve dilemmas (see, for example, Brodie 2013).

2.5.4 Suggestions for Future Work

There are many challenges in adopting new assessment practices and there is no doubt that this is a complex issue. This section has suggested some strategies to support teachers as they adopt new assessment strategies in their classroom, but many issues still remain. Some continued and future areas of research might include:

- How is assessment handled in teacher preparation programs? Are issues about assessment handled by generalists or by those fluent in assessment in mathematics education? What are advantages/disadvantages of having mathematics teacher educators engage teachers in professional development related to assessment?
- While this research discusses supporting teachers who are adopting new practices in assessment, what about those teachers who are reluctant to make those shifts?

3 Summary and Looking Ahead

This monograph has focused on a variety of issues that relate to assessment, both large-scale assessment and classroom assessment. In particular, the monograph has highlighted the following:

- Purposes, traditions, and principles of assessment have been described and summarized.
- Issues related to the design of assessment tasks have been outlined, with principles from various design models compared.
- Assessment has been described in action, including strategies specifically designed for classroom use and issues related to the interactions of large-scale and classroom assessment.
- Issues related to enhancing sound assessment knowledge and practices have been described with examples of potential models for teacher support.
- Questions for future research and discussion at the conference have been articulated within each of the major sections of the monograph.

References

Adler, J. (1998). A language of teaching dilemmas: Unlocking the complex multilingual secondary mathematics classroom. *For the Learning of Mathematics, 18*(1), 24–33.

Baird, J., Hopfenbeck, T. N., Newton, P., Stobart, G., & Steen-Utheim, A. T. (2014). *State of the field review: Assessment and learning. Norwegian Knowledge Centre for Education.* Oxford, England: Oxford University Centre for Educational Assessment.

Barnes, M., Clarke, D., & Stephens, M. (2000). Assessment: The engine of systemic curricular reform? *Journal of Curriculum Studies, 32*(5), 623–650.

Beckers, J. (2002). *Développer et évaluer des compétences à l'école: vers plus d'efficacité et d'équité.* Brussels, Belgium: Labor.

Black, P., & Wiliam, D. (2006). Developing a theory of formative assessment. In J. Gardner (Ed.), *Assessment and learning* (pp. 81–100). London, England: Sage.

Black, P. J., & Wiliam, D. (2009). Developing the theory of formative assessment. *Educational Assessment, Evaluation and Accountability, 21*(1), 5–31.

Black, P., Harrison, C., Lee, C., Marshall, B., & Wiliam, D. (2004). Inside the black box: Assessment for learning in the classroom. *Phi Delta Kappan, 86,* 8–21.

Bleiler, S., Ko, Y. Y., Yee, S. P., & Boyle, J. D. (2015). Communal development and evolution of a course rubric for proof writing. In C. Suurtamm & A. Roth-McDuffie (Eds.), *Annual perspectives in mathematics education: Assessment to enhance teaching and learning* (pp. 97–108). Reston, VA: National Council of Teachers of Mathematics.

Bleiler, S. K., & Thompson, D. R. (2012/2013). Multi-dimensional assessment of the common core. *Teaching Children Mathematics, 19*(5), 292–300.

Bodin, A. (1997). L'évaluation du savoir mathématique. *Recherches En Didactique Des Mathématiques, Savoirs & methods, 17*(1), 49–96.

Boudett, K. P., & Steele, J. L. (2007). *Data wise in action: Stories of schools using data to improve teaching and learning.* Cambridge, MA: Harvard University Press.

Boudett, K. P., City, E. A., & Murnane, R. J. (2008). *Data-wise: A step-by-step guide to using assessment results to improve teaching and learning.* Cambridge, MA: Harvard Education Press.

Brodie, K. (2013). The power of professional learning communities. *Education as Change, 17*(1), 5–18.

Brookhart, S. M. (2003). Developing measurement theory for classroom assessment purposes and uses. *Educational Measurement: Issues and Practice, 22*(4), 5–12.

Brunner, M., Artelt, C., Krauss, S., & Baumert, J. (2007). Coaching for the PISA test. *Learning and Instruction, 17,* 111–122.

Care, E., Griffin, P., Zhang, Z., & Hutchinson, D. (2014). Large-scale testing and its contribution to learning. In C. Wyatt-Smith, V. Klenowski, & P. Colbert (Eds.), *Designing assessment for quality learning* (pp. 55–72). Dordrecht, The Netherlands: Springer.

Cheang, W. K., Teo, K. M., & Zhao, D. (2012). Assessing mathematical competences using disciplinary tasks. In *Proceedings of the 12th International Congress on Mathematical Education: Topic Study Group 33* (pp. 6504–6513). Seoul, Korea.

Crespo, S., & Rigelman, N. (2015). Introduction to Part III. In C. Suurtamm & A. Roth McDuffie (Eds.), *Annual perspectives in mathematics education: Assessment to enhance teaching and learning* (pp. 119–121). Reston, VA: National Council of Teachers of Mathematics.

De Ketele, J.-M. (1989). L'évaluation de la productivité des institutions d'éducation. *Cahier de la Fondation Universitaire; Université et Société, le rendement de l'enseignement universitaire, 3*, 73–83.

De Lange, J. (1992). Critical factors for real changes in mathematics learning. In G. C. Leder (Ed.), *Assessment and learning of mathematics* (pp. 305–329). Hawthorn, Victoria: Australian Council for Educational Research.

De Lange, J. (2007). Large-scale assessment and mathematics education. In F. K. Lester, Jr. (Ed.), *Second handbook of research on mathematics teaching and learning* (pp. 1111–1142). Charlotte, NC: Information Age Publishing.

Dempsey, K., Beesley, A. D., Clark, T. F., & Tweed, A. (2015). Authentic student work samples support formative assessment in middle school. In C. Suurtamm & A. Roth-McDuffie (Eds.), *Annual perspectives in mathematics education: Assessment to enhance teaching and learning* (pp. 157–165). Reston, VA: National Council of Teachers of Mathematics.

Dyrvold, A., Bergqvist, E., & Österholm, M. (2015). Uncommon vocabulary in mathematical tasks in relation to demand of reading ability and solution frequency. *Nordic Studies in Mathematics Education, 20*(1), 5–31.

Duncan, C. R., & Noonan, B. (2007). Factors affecting teachers' grading and assessment practices. *Alberta Journal of Educational Research, 53*(1), 1–21.

Earl, L., & Timperley, H. (2014). Challenging conceptions of assessment. In C. Wyatt-Smith, V. Klenowski, & P. Colbert (Eds.), *Designing assessment for quality learning* (pp. 325–336). Dordrecht, The Netherlands: Springer.

Earl, L., & Torrance, N. (2000). Embedding accountability and improvement into large-scale assessment: What difference does it make? *Peabody Journal of Education, 75*(4), 114–141.

Elrod, M. J., & Strayer, J. F. (2015). Using an observational rubric to facilitate change in undergraduate classroom norms. In C. Suurtamm & A. Roth McDuffie (Eds.), *Annual perspectives in mathematics education: Assessment to enhance teaching and learning* (pp. 87–96). Reston, VA: National Council of Teachers of Mathematics.

Engelsen, K. S., & Smith, A. (2014). Assessment literacy. In C. Wyatt-Smith, V. Klenowski, & P. Colbert (Eds.), *Designing assessment for quality learning* (pp. 91–107). Dordrecht, The Netherlands: Springer.

Even, R. (2005). Using assessment to inform instructional decisions: How hard can it be? *Mathematics Education Research Journal, 17*(3), 45–61.

Fan, L. (2011). Implementing self-assessment to develop reflective teaching and learning in mathematics. In B. Kaur & K. Y. Wong (Eds.), *Assessment in the mathematics classroom: 2011 Association of Mathematics Educators Yearbook* (pp. 275–290). Singapore: World Scientific Publishing.

Gardner, J. (Ed.). (2006). *Assessment and learning*. Thousand Oaks, CA: Sage.

Gipps, C. (1994). *Beyond testing: Towards a theory of educational assessment*. London, England: Falmer Press.

Glaser, R., & Silver, E. A. (1994). Assessment, testing, and instruction: Retrospect and prospect. In L. Darling-Hammond (Ed.), *Review of research in education* (Vol. 20, pp. 393–419). Washington, DC: American Educational Research Association.

Graue, M. E., & Smith, S. Z. (1996). Shaping assessment through instructional innovation. *Journal of Mathematical Behavior, 15*(2), 113–136.

Hunsader, P. D., Thompson, D. R., Zorin, B., Mohn, A. L., Zakrzewski, J., Karadeniz, I., Fisher, E. C., & MacDonald, G. (2014). Assessments accompanying published textbooks: The extent to which mathematical processes are evident. *ZDM: The International Journal on Mathematics Education, 46*(5), 797–813. doi:10.1007/s11858-014-0570-6.

Hunsader, P. D., Thompson, D. R., & Zorin, B. (2015a). Developing teachers' ability to be critical consumers of assessments. In C. Suurtamm & A. Roth-McDuffie (Eds.), *Annual perspectives in mathematics education: Assessment to enhance teaching and learning* (pp. 123–132). Reston, VA: National Council of Teachers of Mathematics.

Hunsader, P. D., Zorin, B., & Thompson, D. R. (2015b). Enhancing teachers' assessment of mathematical processes through test analysis in university courses. *Mathematics Teacher Educator, 4*(1), 71–92.

Intanate, N. (2012). Exploring mathematics student teachers' practices on classroom assessment through professional development by lesson study and open approach. In Poster presented at *the 12th International Congress on Mathematical Education: Topic Study Group 33*. Seoul, Korea.

Jin, H., & Wong, K. Y. (2011). Assessing conceptual understanding in mathematics with concept mapping. In B. Kaur & K. Y. Wong (Eds.), *Assessment in the mathematics classroom: 2011 Association of Mathematics Educators Yearbook* (pp. 67–90). Singapore: World Scientific Publishing.

Joint Committee on Standards for Educational Evaluation. (2003). *The student evaluation standards: How to improve evaluations of students.* Thousand Oaks, CA: Corwin Press.

Kaur, B., & Chan, C. M. E. (2011). Using journal writing to empower learning. In B. Kaur & K. Y. Wong (Eds.), *Assessment in the mathematics classroom: 2011 Association of Mathematics Educators Yearbook* (pp. 91–112). Singapore: World Scientific Publishing.

Kenney, P. A., & Silver, E. A. (1993). Student self-assessment in mathematics. In N. L. Webb & A. F. Coxford (Eds.), *Assessment in the mathematics classroom, K-12 [1993 Yearbook of the National Council of Teachers of Mathematics]* (pp. 229–238). Reston, VA: National Council of Teachers of Mathematics.

Kim, M.-J., & Lehrer, R. (2015). Using learning progressions to design instructional trajectories. In C. Suurtamm & A. Roth McDuffie (Eds.), *Annual perspectives in mathematics education: Assessment to enhance teaching and learning* (pp. 27–38). Reston, VA: National Council of Teachers of Mathematics.

Kim, R. Y., Kim, K. Y., Lee, M. H., Jeon, J. H., & Park, J. W. (2012). The challenges and issues regarding extended constructed-response questions: Korean teachers' perspective. In *Proceedings of the 12th International Congress on Mathematical Education: Topic Study Group 33* (pp. 6631–6640). Seoul, Korea.

Klenowski, V. (2009). Assessment for Learning revisited: An Asia-Pacific perspective. *Assessment in Education: Principles, Policy, and Practice, 16*(3), 263–268.

Klieme, E., Avenarius, H., Blum, W., Döbrich, P., Gruber, H., Prenzel, M., et al. (2004). *The development of national educational standards: An expertise.* Berlin, Germany: Federal Ministry of Education and Research.

Klinger, D. A., McDivitt, P. R., Howard, B. B., Munoz, M. A., Rogers, W. T., & Wylie, E. C. (2015). *The classroom assessment standards for preK-12 teachers.* Kindle Direct Press.

Koch, M. J. (2013). The multiple-use of accountability assessments: Implications for the process of validation. *Educational Measurement: Issues and Practice, 32*(4), 2–15.

Krzywacki, H., Koistinen, L., & Lavonen, J. (2012). Assessment in Finnish mathematics education: Various ways, various needs. In *Proceedings of the 12th International Congress on Mathematical Education: Topic Study Group 33* (pp. 6661–6670). Seoul, Korea.

Leahy, S., Lyon, C., Thompson, M., & Wiliam, D. (2005). Classroom assessment: Minute by minute, day by day. *Educational Leadership, 63*(3), 19–24.

Lee, J-E., Turner, H., Ansara, C., Zablocki, J., Hincks, C., & Hanley, V. (2015). Practicing questioning in a professional learning community: A hub of classroom assessment. In C. Suurtamm & A. Roth McDuffie (Eds.), *Annual perspectives in mathematics education: Assessment to enhance teaching and learning* (pp. 133–143). Reston, VA: National Council of Teachers of Mathematics.

Levin, T., & Shohamy, E. (2008). Achievement of immigrant students in mathematics and academic Hebrew in Israeli schools: A large-scale evaluation study. *Studies in Educational Evaluation, 34*(1), 1–14.

Lund, A. (2008). Assessment made visible: Individual and collective practices. *Mind, Culture, and Activity, 15*, 32–51.

Marynowski, R. (2015). Formative assessment strategies in the secondary mathematics classroom. In C. Suurtamm & A. Roth McDuffie (Eds.), *Annual perspectives in mathematics education: Assessment to enhance teaching and learning* (pp. 167–173). Reston, VA: National Council of Teachers of Mathematics.

Mathematical Sciences Education Board (MSEB), & National Research Council (NRC). (1993). *Measuring up: Prototypes for mathematics assessment.* Washington, DC: National Academy Press.

McTighe, J. (2015). What is a performance task? Retrieved from http://www.performancetask. com/what-is-a-performance-task/.

Messick, S. (1989). Validity. In R. L. Linn (Ed.), *Educational measurement* (3rd ed., pp. 13–103). New York, NY: Macmillan.

Mok, M. M. C. (2011). The assessment for, of, and as learning in mathematics: The application of SLOA. In B. Kaur & K. Y. Wong (Eds.), *Assessment in the mathematics classroom: 2011 Association of Mathematics Educators Yearbook* (pp. 187–215). Singapore: World Scientific Publishing.

National Council of Teachers of Mathematics (NCTM). (1995). *Assessment standards for school mathematics.* Reston, VA: Author.

National Council of Teachers of Mathematics (NCTM). (2000). *Principles and standards for school mathematics.* Reston, VA: Author.

National Council of Teachers of Mathematics (NCTM). (2014). *Principles to action: Ensuring mathematical success for all.* Reston, VA: Author.

National Research Council. (2001a) *Adding it up: Helping children learn mathematics.* J. Kilpatrick, J. Swafford, & B. Findell (Eds.), Mathematics Learning Study Committee, Center for Education, Division of Behavior and Social Sciences and Education. Washington, DC: National Academy Press.

National Research Council. (2001b). *Classroom assessment and the national science education standards.* J. M. Atkin, P. Black, & J. Coffey (Eds.), Committee on Classroom Assessment and the National Science Education Standards. Washington, DC: National Academy Press.

Newton, P. E. (2007). Clarifying the purposes of educational assessment. *Assessment in Education: Principles, Policy and Practice, 14*(2), 149–170.

Obersteiner, A., Moll, G., Reiss, K., & Pant, H. A. (2015). *Whole number arithmetic—competency models and individual development.* Paper delivered at ICMI Study 23, Macau. (pp. 235–242).

Organization for Economic Cooperation and Development (OECD). (2009). *PISA 2006 technical report.* Paris, France: Author.

Osterlind, S. J. (1998). *Constructing test items: Multiple-choice, constructed-response, performance and other formats.* Dordrecht, The Netherlands: Kluwer Academic Publishers.

Paek, P. L. (2012). Using learning trajectories in large-scale mathematics assessments. In *Proceedings of the 12th International Congress on Mathematical Education: Topic Study Group 33* (pp. 6711–6720). Seoul, Korea.

Parke, C., Lane, S., Silver, E. A., & Magone, M. (2003). *Using assessment to improve mathematics teaching and learning: Suggested activities using QUASAR tasks, scoring criteria, and student work.* Reston, VA: National Council of Teachers of Mathematics.

Pellegrino, J. W., Chudowsky, N., & Glaser, R. (Eds.). (2001). *Knowing what students know: The science of design and educational assessment.* Washington, DC: National Academy Press.

Perrenoud, P. (1997). *Construire des compétences dès l'école.* Paris, France: ESF Editeur.

Pólya, G. (1945). *How to solve it: A new aspect of mathematical method.* Princeton, NJ: Princeton University Press.

Pryor, J., & Crossouard, B. (2008). A socio-cultural theorisation of formative assessment. *Oxford Review of Education, 34*(1), 1–20.

Rankin, J. G. (2015). Guidelines for analyzing assessment data to inform instruction. In C. Suurtamm & A. Roth-McDuffie (Eds.), *Annual perspectives in mathematics education:*

Assessment to enhance learning and teaching (pp. 191–198). Reston, VA: National Council of Teachers of Mathematics.

Rey, B., Carette, V., Defrance, A., & Kahn, S. (2003). *Les compétences à l'école: apprentissage et évaluation*. Brussels, Belgium: De Boeck.

Sayac, N., & Grapin, N. (2015). Evaluation externe et didactique des mathématiques: un regard croisé nécessaire et constructif. *Recherches en didactique des mathématiques, 35*(1), 101–126.

Scherrer, J. (2015). Learning, teaching, and assessing the standards for mathematical practice. In C. Suurtamm & A. Roth-McDuffie (Eds.), *Annual perspectives in mathematics education: Assessment to enhance learning and teaching* (pp. 199–208). Reston, VA: National Council of Teachers of Mathematics.

Schoenfeld, A. (1992). Learning to think mathematically: Problem solving, metacognition, and sense making in mathematics. In D. A. Grouws (Ed.), *Handbook of research on mathematics teaching and learning* (pp. 334–370). New York, NY: MacMillan Publishing.

Schukajlow, S., Kolter, J., & Blum, W. (2015). Scaffolding mathematical modelling with a solution plan. *ZDM: International Journal on Mathematics Education, 47*(7), 1241–1254.

Schukajlow, S., Krug, A., & Rakoczy, K. (2015b). Effects of prompting multiple solutions for modelling problems on students' performance. *Educational Studies in Mathematics, 89*(3), 393–417. doi:10.1007/s10649-015-9608-0.

Sears, R., Karadeniz, I., Butler, K., & Pettey, D. (2015). Are standards for mathematical practice overlooked in geometry textbooks' chapter tests? In C. Suurtamm & A. Roth McDuffie (Eds.), *Annual perspectives in mathematics education: Assessment to enhance teaching and learning* (pp. 75–86). Reston, VA: National Council of Teachers of Mathematics.

Semena, S., & Santos, L. (2012). Towards the students' appropriation of assessment criteria. Poster presented at *12th International Congress on Mathematical Education: Topic Study Group 33*, Seoul, Korea.

Shalem, Y., Sapire, I., & Huntley, B. (2012). How curriculum mapping of large-scale assessments can benefit mathematics teachers. In *Proceedings of the 12th International Congress on Mathematical Education: Topic Study Group 33* (pp. 6601–6610). Seoul, Korea.

Shepard, L. A. (2000). The role of assessment in a learning culture. *Educational Researcher, 29*(7), 4–14.

Shepard, L. A. (2001). The role of classroom assessment in teaching and learning. In V. Richardson (Ed.), *The handbook of research on teaching* (4th ed., pp. 1066–1101). Washington, DC: American Educational Research Association.

Shimizu, Y. (2011). Building bridges between large-scale external assessment and mathematics classrooms: A Japanese perspective. In B. Kaur & K. Y. Wong (Eds.), *Assessment in the mathematics classroom: 2011 Association of Mathematics Educators Yearbook* (pp. 217–235). Singapore: World Scientific Publishing.

Silver, E. A., Ghousseini, H., Gosen, D., Charalambous, C., & Strawhun, B. T. F. (2005). Moving from rhetoric to praxis: Issues faced by teachers in having students consider multiple solutions for problems in the mathematics classroom. *Journal of Mathematical Behavior, 24*, 287–301.

Smit, R., & Birri, Th. (2014). Assuring the quality of standards-oriented classroom assessment with rubrics for complex competencies. *Studies in Educational Evaluation, 43*, 5–13.

Stobart, G. (2008). *Testing times: The uses and abuses of assessment*. London, England: Routledge.

Suurtamm, C., & Koch, M. J. (2014). Navigating dilemmas in transforming assessment practices: Experiences of mathematics teachers in Ontario, Canada. *Educational Assessment, Evaluation and Accountability, 26*(3), 263–287.

Suurtamm, C., Koch, M., & Arden, A. (2010). Teachers' assessment practices in mathematics: Classrooms in the context of reform. *Assessment in Education: Principles, Policy, and Practice, 17*(4), 399–417.

Suurtamm, C., & Neubrand, M. (2015). Assessment and testing in mathematics education. In S. J. Cho (Ed.), *The Proceedings of the 12th International Congress on Mathematical Education* (pp. 557–562). Dordrecht, The Netherlands: Springer.

Swan, M., & Burkhardt, H. (2012). A designer speaks: Designing assessment of performance in mathematics. *Educational Designer: Journal of the International Society for Design and Development in Education, 2*(5), 1–41. http://www.educationaldesigner.org/ed/volume2/issue5/article19.

Tillema, H. H. (2014). Student involvement in assessment of their learning. In C. Wyatt-Smith, V. Klenowski, & P. Colbert (Eds.), *Designing assessment for quality learning* (pp. 39–54). Dordrecht, The Netherlands: Springer.

Tillema, H. H., & Kremer-Hayon, L. (2005). Facing dilemmas: Teacher educators' ways to construct a pedagogy of teacher education. *Teaching in Higher Education, 10*(2), 207–221.

Timperley, H. (2014). Using assessment information for professional learning. In C. Wyatt-Smith, V. Klenowski, & P. Colbert (Eds.), *Designing assessment for quality learning* (pp. 137–150). Dordrecht, The Netherlands: Springer.

Thompson, D. R., & Kaur, B. (2011). Using a multi-dimensional approach to understanding to assess students' mathematical knowledge. In B. Kaur & K. Y. Wong (Eds.), *Assessment in the mathematics classroom: 2011 Association of Mathematics Educators Yearbook* (pp. 17–32). Singapore: World Scientific Publishing.

Toh, T. L., Quek, K. S., Leong, Y. H., Dindyal, J., & Tay, E. G. (2011). Assessing problem solving in the mathematics curriculum: A new approach. In B. Kaur & K. Y. Wong (Eds.), *Assessment in the mathematics classroom: 2011 Association of Mathematics Educators Yearbook* (pp. 33–66). Singapore: World Scientific Publishing.

Ufer, S., Reiss, K., & Mehringer, V. (2013). Sprachstand, soziale herkunft und bilingualität: Effekte auf facetten mathematischer kompetenz. In M. Becker-Mrotzek, K. Schramm, E. Thürmann, & H. J. Vollmer (Eds.), *Sprache im fach: Sprachlichkeit und fachliches lerner* (pp. 167–184). Münster, Germany: Waxmann. (*Language, skills, social background and migration: Effects on facets of mathematics skills*).

Van den Heuvel-Panhuizen, M., & Becker, J. (2003). Towards a didactic model for assessment design in mathematics education. In A. J. Bishop, M. A. Clements, C. Keitel, J. Kilpatrick, & F. K. S. Leung (Eds.), *Second international handbook of mathematics education* (pp. 686–716). Dordrecht, The Netherlands: Kluwer Academic Publishers.

Vescio, V., Ross, D., & Adams, A. (2008). A review of research on the impact of professional learning communities on teaching practice and student learning. *Teaching and Teacher Education, 24*(1), 80–91.

Vos, P. (2005). Measuring mathematics achievement: A need for quantitative methodology literacy. In J. Adler & M. Kazima (Eds.), *Proceedings of the 1st African Regional Congress of the International Commission on Mathematical Instruction*. Johannesburg, South Africa: University of the Witwatersrand.

Vos, P., & Kuiper, W. A. J. M. (2003). Predecessor items and performance level. *Studies in Educational Evaluation, 29*, 191–206.

Walcott, C. Y., Hudson, R., Mohr, D., & Essex, N. K. (2015). What NAEP tells us about the relationship between classroom assessment practices and student achievement in mathematics. In C. Suurtamm & A. Roth McDuffie (Eds.), *Annual perspectives in mathematics education: Assessment to enhance teaching and learning* (pp. 179–190). Reston, VA: National Council of Teachers of Mathematics.

Watson, A. (2006). Some difficulties in informal assessment in mathematics. *Assessment in Education, 13*(3), 289–303.

Watson, A., & Ohtani, M. (Eds.). (2015a). *Task design in mathematics education: An ICMI Study 22*. Heidelberg, Germany: Springer.

Webb, D. C. (2012). Teacher change in classroom assessment: The role of teacher content knowledge in the design and use of productive classroom assessment. In *Proceedings of the 12th International Congress on Mathematical Education: Topic Study Group 33* (pp. 6773–6782). Seoul, Korea.

Webb, N. M. (1993). Collaborative group versus individual assessment in mathematics: Processes and outcomes. *Educational Assessment, 1*(2), 131–152.

Webb, N. L. (2002). *Assessment literacy in a standards-based urban education setting*. Paper presented to the American Educational Research Association Annual Meeting, New Orleans, Louisiana.

Wiliam, D. (2007). Keeping learning on track: Classroom assessment and the regulation of learning. In F. K. Lester, Jr. (Ed.), *Second handbook of research on mathematics teaching and learning* (pp. 1053–1098). Charlotte, NC: Information Age Publishing.

Wiliam, D. (2011a). *Embedded formative assessment: Practical strategies and tools for K-12 teachers*. Bloomington, IN: Solution Tree Press.

Wiliam, D. (2011b). What is assessment for learning? *Studies in Educational Evaluation, 37*, 3–14.

Wiliam, D. (2015). Assessment: A powerful focus for the improvement of mathematics instruction. In C. Suurtamm & A. Roth McDuffie (Eds.), *Annual perspectives in mathematics education: Assessment to enhance teaching and learning* (pp. 247–254). Reston, VA: National Council of Teachers of Mathematics.

Wilson, S. M., & Kenney, P. A. (2003). Classroom and large-scale assessment. In J. Kilpatrick, W. G. Martin, & D. Schifter (Eds.), *A research companion to principles and standards for school mathematics* (pp. 53–67). Reston, VA: National Council of Teachers of Mathematics.

Windschitl, M. (2002). Framing constructivism in practice as the negotiation of dilemmas: An analysis of the conceptual, pedagogical, cultural, and political challenges facing teachers. *Review of Educational Research, 72*(2), 131–175.

Wyatt-Smith, C., Klenowski, V., & Colbert, P. (2014). Assessment understood as enabling. In C. Wyatt-Smith, V. Klenowski, & P. Colbert (Eds.), *Designing assessment for quality learning* (pp. 1–20). Dordrecht, The Netherlands: Springer.

Yamamoto, S. (2012). Finding descriptive feedback in Japanese 1st grade mathematics class. Poster presented at *the 12th International Congress on Mathematical Education: Topic Study Group 33*. Seoul, Korea.

Young-Loveridge, J., & Bicknell, B. (2015). Using task-based interviews to assess early understanding of number. In C. Suurtamm & A. Roth McDuffie (Eds.), *Annual perspectives in mathematics education: Assessment to enhance teaching and learning* (pp. 67–74). Reston, VA: National Council of Teachers of Mathematics.

Further Reading

Collins, A. M. (Ed.). (2011). *Using classroom assessment to improve student learning*. Reston, VA: National Council of Teachers of Mathematics.
This book focuses on issues of classroom formative assessment, with sample tasks and student work from middle grades classrooms to illustrate what teachers might learn about students' thinking from such tasks.

Kohn, A. (2000). *The case against standardized testing: Raising the scores, ruining the schools*. Portsmouth, NH: Heinemann.
Although not focused specifically on mathematics, the author addresses many of the issues and criticisms against standardized testing with responses to counter rationales provided by policy makers for such assessments. Even though the focus is set within the context of the USA, the issues may be relevant to those in other countries in which high stakes assessments are being used to judge schools.

Kopriva, R., & Saez, S. (1997). *Guide to scoring LEP student responses to open-ended mathematics items*. Washington, DC: Council of Chief State School Officers. (ERIC ED452704).
This document highlights issues of scoring open-ended assessment items from students whose mother tongue is not the language in which the assessment is administered, and considers errors often made at different levels of language proficiency.

Lane, S., & Silver, E. A. (1999). Fairness and equity in measuring student learning using a mathematics performance assessment: Results from the QUASAR project. In A. L. Nettles & M. T. Nettles (Eds.), *Measuring up: Challenges minorities face in educational assessment* (pp. 97–120). Boston, MA: Kluwer.
This paper describes how issues of equity and fairness were addressed in the design of a large-scale mathematics assessment intended to measure students' mathematics proficiency, with an emphasis on conceptual understanding and problem solving.

Magone, M., Cai, J., Silver, E. A., & Wang, N. (1994). Validating the cognitive complexity and content quality of a mathematics performance assessment. *International Journal of Educational Research, 21*(3), 317–340.
This paper describes how the cognitive complexity of assessment tasks was addressed in the design of a large-scale mathematics assessment intended to measure students' mathematics proficiency, with an emphasis on conceptual understanding and problem solving.

OECD. (2011). *Strong performers and successful reformers in education: Lessons from PISA for the United States*. OEDC Publishing.
This document looks at lessons from educational systems in Canada, Shanghai, Hong Kong, Finland, Japan, Singapore, Brazil, Germany, England, and Poland based on PISA results and draws lessons from those countries, specifically for the United States.

Silver, E. A. (1994). Mathematical thinking and reasoning for all students: Moving from rhetoric to reality. In D. F. Robitaille, D. H. Wheeler, & C. Kieran (Eds.), *Selected lectures from the 7th International Congress on Mathematical Education* (pp. 311–326). Sainte-Foy, Quebec: Les Presses De L'Université Laval.
This paper describes how issues of the interplay of assessment and instruction were addressed in the design of a large-scale mathematics instructional reform project intended to increase the mathematics proficiency of students attending schools in economically disadvantaged communities in the USA, with an emphasis on conceptual understanding and problem solving.

Silver, E. A., Alacaci, C., & Stylianou, D. A. (2000). Students' performance on extended constructed-response tasks. In E. A. Silver & P. A. Kenney (Eds.), *Results from the Seventh Mathematics Assessment of the National Assessment of Educational Progress* (pp. 301–341). Reston, VA: National Council of Teachers of Mathematics.
This paper reports a secondary analysis of students' responses to complex tasks that were part of a large-scale assessment in the USA. The analysis reveals information about students' solution strategies, representational approaches, and common errors.

Silver, E. A., & Kenney, P. A. (Eds.). (2000). *Results from the Seventh Mathematics Assessment of the National Assessment of Educational Progress*. Reston, VA: National Council of Teachers of Mathematics.
The chapters in the book report secondary examinations of results from a large-scale assessment in the USA. Results are discussed in relation to related research and implications are drawn to assist teachers, teacher educators, and curriculum developers to use the assessment results productively.

Silver, E. A., & Lane, S. (1993). Assessment in the context of mathematics instruction reform: The design of assessment in the QUASAR project. In M. Niss (Ed.), *Assessment in mathematics education and its effects* (pp. 59–70). London, England: Kluwer.
This paper describes how a large-scale assessment was developed in the context of a mathematics instructional reform project intended to increase the mathematics proficiency of students attending schools in economically disadvantaged communities in the USA, with an emphasis on conceptual understanding and problem solving.

Silver, E. A., & Smith, M. S. (2015). Integrating powerful practices: Formative assessment and cognitively demanding mathematics tasks. In C. Suurtamm & A. Roth McDuffie (Eds.), *Annual perspectives in mathematics education: Assessment to enhance teaching and learning* (pp. 5–14). Reston, VA: National Council of Teachers of Mathematics.
This paper discusses and illustrates, through a classroom lesson scenario, the synergistic interplay of formative assessment and the use of cognitively demanding mathematics tasks in classroom instruction.

Smith, G., Wood, L., Coupland, M., Stephenson, B., Crawford, K., & Ball, G. (1996). Constructing mathematical examinations to assess a range of knowledge and skills. *International Journal of Mathematical Education in Science and Technology, 27*(1), 65–77. This paper introduces a taxonomy that can be used to design and classify assessment tasks in undergraduate mathematics classes that focus not only on factual knowledge but on important processes as well as helping students reflect on learning.

Stacey, K. (2016). Mathematics curriculum, assessment and teaching for living in the digital world: Computational tools in high stakes assessment. In M. Bates & Z. Usiskin (Eds.), *Digital curricula in school mathematics* (pp. 251–270). Charlotte, NC: Information Age Publishing. This paper describes tools used in high school assessments in Australia, particularly assessments using computer algebra systems, and provides some research results from studies conducted with students having access to CAS as part of normal school instruction.

Stacey, K., & Turner, R. (Eds.). (2015). *Assessing mathematical literacy: The PISA experience.* Heidelberg, Germany: Springer. This book discusses the conceptualization, design, development, background and impact of the mathematics assessment for the OECD Programme for International Student Assessment (PISA). The book includes a number of papers that focus on questions important for mathematics education, such as the analysis of effects of competences on item difficulties.

Stacey, K., & Wiliam, D. (2012). Technology and assessment in mathematics. In M. A. K. Clements, A. Bishop, C. Keitel-Kreidt, J. Kilpatrick, & F. Koon-Shing Leung (Eds.), *Third international handbook of mathematics education* (pp. 721–751). New York, NY: Springer Science & Business Media. This chapter presents a detailed discussion of the constraints and affordances with respect to assessment that need to be considered as technology becomes an increasing component of mathematics teaching and learning.

Stylianou, D. A., Kenney, P. A., Silver, E. A., & Alacaci, C. (2000). Gaining insight into students' thinking through assessment tasks. *Mathematics Teaching in the Middle School, 6*, 136–144. This paper illustrates how assessment tasks drawn from large-scale assessment can be used by teachers to gain access to students' mathematical thinking and reasoning.

Watson, A., & Ohtani, M. (Eds.). (2015b). *Task design in mathematics education: An ICMI Study 22.* Heidelberg, Germany: Springer. The chapters in this book focus on various aspects of task design based on an ICMI study.

Wilson, L. D. (2007). High-stakes testing in mathematics. In F. K. Lester, Jr. (Ed.), *Second handbook of research on mathematics teaching and learning* (pp. 1099–1110). Charlotte, NC: Information Age Publishing. This chapter reviews research related to high-stakes testing and highlights many of the technical aspects of such assessments.

Wyatt-Smith, C., Klenowski, V., & Colbert, P. (Eds.). (2014b). *Designing assessment for quality learning.* Dordrecht, The Netherlands: Springer. This chaptered book provides a range of current perspectives and issues on assessment.

www.ingramcontent.com/pod-product-compliance
Ingram Content Group UK Ltd.
Pitfield, Milton Keynes, MK11 3LW, UK
UKHW020216231225
466357UK00011B/180